REPORT OF THE COMMITTEE

ON

COINAGE, WEIGHTS, AND MEASURES,

TO THE

HOUSE OF REPRESENTATIVES, U. S.

MAY 17, 1866.

WASHINGTON:
GOVERNMENT PRINTING OFFICE.
1866.

39TH CONGRESS, 1st Session. } HOUSE OF REPRESENTATIVES. { REPORT No. 62.

COINAGE, WEIGHTS, AND MEASURES.

[To accompany bills H. R. Nos. 596 and 597, and H. Res. No. 141.]

MAY 17, 1866.—Ordered to be printed.

Mr. KASSON, from the Committee on Coinage, Weights, and Measures, made the following

REPORT.

In considering the general subject of a uniform system of coinage, weights, and measures, your committee have had before them—

First. That part of the message of the President and the accompanying documents relating to these subjects.

Second. The report of the National Academy of Sciences, embracing their resolutions approving the metric decimal system of weights and measures.

Third. The report of the United States commissioner to the statistical congress at Berlin.

Fourth. Various memorials of universities and colleges of the United States urging a uniform system of weights and measures, also invariably commending the metric decimal system.

Fifth. The petition of the mayor, judges, and citizens of Baltimore praying for the adoption of the metric system of weights and measures.

Sixth. Several memorials of citizens in different parts of the United States in behalf of the same object.

Seventh. The bill H. R. No. 252, referred to them, and proposing the compulsory and exclusive use, after a limited period, of the metric system.

In addition to the documents and papers referred to them, and in the absence of authority to send for persons and papers, which they did not regard as indispensable to a proper investigation of the subjects at this time, they have examined the whole history of the efforts made in this country since the adoption of the Constitution to substitute for our imperfect and incongruous system of weights and measures, a system at once simple, complete, uniform, and decimal in its relations. The result of that examination is embodied in this report. They have also carefully examined the testimony taken before the select parliamentary committee on this subject in England—testimony very complete, and almost exhaustive of both facts and reasoning—touching the various phases of the questions involved. To these investigations they have added inquiries into the public action of other countries with which we have established commercial relations, on both the European and American continents. They also received the assistance of those distinguished members of the National Academy of Sciences who constituted the special committee of that learned society having charge of these subjects, and particularly of Professor Newton, of that committee, whose efforts in aid of their purposes have been patient and persevering.

The troubled condition of the United States, and the consequent extraordinary labors thrown upon the 38th Congress, prevented your committee from then undertaking that thorough examination which the importance of the questions demanded. They now, however, submit their report and the accompanying bills, as indicating the conclusions to which they have unanimously come at this period of their deliberations. They do not doubt that a subsequent Congress will be prepared to go further, and will enable the republic to lead, rather than to follow, the action of other commercial and intelligent nations in the complete establishment of this most urgently demanded reform. It is an obligation we owe not only to our present convenience, but also to posterity, to whose benefit all sound reforms invariably tend.

COINAGE.

In respect to the gold and silver coins of the United States no specific change can, with propriety, be recommended for immediate adoption.

The United States early established (July 6, 1785) the decimal system in its application to money, and as a consequence of it, have now a simple, convenient, and admirable measure of values. It only remains to be considered how a common standard of international values, for the use of all civilized and commercial nations, may be most conveniently established.

In this connexion three questions arise:

First. Should a unit entirely new be established? or,

Second. Should the established unit of some one nation, now in use, be adopted by all other nations?

Third. If so, which possesses the greatest advantages?

The advantage of the decimal system is now universally conceded among commercial nations. No country is more ready to concede its superiority than England, which has hitherto failed to adopt it. It is understood that the Bank of England, and some of her great railroad corporations have been compelled to adopt the decimal system in the keeping of accounts. The government has also created a new coin in order to obtain the tenth of a pound.

The objections to the creation of an entirely new unit of value are evidently irresistible, if any existing unit in national use meets the conditions of convenience as a common standard, and of decimal computation; for the people of all nations will be subject to the inconvenience of the adoption of the wholly new system, while the adoption of a decimal system now in use in any one or more nations would relieve at least a portiom of the people from the inconveniences attending the change, and would find the people of all nations at least partially acquainted with it at the beginning. This fact, itself, should justify some sacrifice of national *amour propre* for the general good.

In 1856 (August 15) Congress by a joint resolution directed the Secretary of the Treasury to appoint a commissioner to confer with the proper functionaries in Great Britain in relation to some plan of so mutually arranging, on a decimal basis, the coinage of the two countries, that the respective units should be thereafter easily and exactly commensurable. The Committee on Finance in the Senate, in reporting the resolution, remark that no measure can be readily suggested whose realization would mark a more decided epoch in the history of commerce. Under this resolution a very competent gentleman was appointed the commissioner, and on the 6th of January, 1859, his report was communicated to Cougress. (Ex. Doc. H. R. No. 36.) Although the British government were not prepared themselves to take the initiative with reference to a project which could not be carried out by it without parliamentary sanction, they were prepared to consider and confer with respect to any proposal that the commissioner might be instructed to make in behalf of the government of the United States.

This result was merely preliminary, but perhaps all that could have been attained under the limited instructions given to the commissioner. But this beginning was not followed up, and there seems to have been no further prosecution of the negotiations.

In his annual report to Congress in December, 1862, the very able Secretary of the Treasury, (now Chief Justice of the United States,) invited the attention of Congress to the present favorable occasion for securing harmony between our own coinage and that of Great Britain. He said:

"In his last report, the Secretary took occasion to invite the attention of Congress to the importance of uniform weights, measures, and coins, and to the worth of the decimal system in the commerce of the world. He now ventures to suggest that the present demonetization of gold may well be availed of for the purpose of taking one considerable step towards these great ends. If the half eagle of the Union be made of equal weight and fineness with the gold sovereign of Great Britain no sensible injury could possibly arise from the change; while, on the resumption of specie payments, its great advantage would be felt in the equalization of exchange and the convenience of commerce. This act of the United States, moreover, might be followed by the adoption by Great Britain of the federal decimal divisions of the coin, and thus a most important advance might be secured towards an international coinage with values decimally expressed."

At the international congress of Berlin, the transactions of which were reported by the United States commissioner, and submitted to Congress, it was resolved as follows:

"First. That the congress recommends that the existing units of money be reduced to a small number; that each unit should be, as far as possible, decimally subdivided; that the coins in use should all be expressed in weights of the metrical system, and should all be of the same degree of fineness, viz: nine-tenths fine, and one-tenth alloy.

"Second. That the different governments be invited to send to a special congress delegates authorized to consider and report what should be the relative weights in the metrical system of gold and silver coins, and to arrange the details by which the monetary system of different countries may be fixed according to the terms of the preceding propositions."

The occasion of the World's Exposition of Industry at Paris in 1867 will furnish the proper opportunity for a free conference between the authorized commissioners of different governments as to the best means of establishing a uniform system of coinage for the common use of the nations of the world. It is to be hoped that the government of the United States will be represented by a commissioner whom it may be authorized to delegate, with special reference to the accomplishment of this great object.

The only interest of any nation that could possibly be injuriously affected by the establishment of this uniformity is that of the money-changers—an interest which contributes little to the public welfare, while, by diversity of coinage and of values, it adds largely to its private accumulations.

The only indispensable condition of this uniformity of value is, that in the standard unit, with its divisions and multiples used in commerce, there shall be in all countries an equal amount of gold (or silver,) with fixed proportion of alloy. Each nation will retain its own devices and legends, and other national peculiarities of mintage. A common name for the standard unit would be desirable, but not essential. The presence of a given amount of precious metal, mixed with a given amount of alloy, is the only absolute prerequisite for the establishment of international uniformity in coinage. The dollar of the United States, four shillings of England, and five francs of France are of approximate value. Several nations of Europe have adopted, under other names, the coinage of France, making it of equal value. The general par value of shares in railroad and other corporations on the continent of Europe, as well as in England and the United States, is one hundred dollars or its approximate equivalent in the money of the different nations. This, of itself, would seem to be a concession of the value which should constitute the standard unit of money. The

United States are now in a favorable condition to yield, with little inconvenience, to a variation in the essential value of their dollar, if it should become necessary, their coin being now withdrawn from general circulation. No opportunity so auspicious for effecting any needed change in quantity of gold or silver, and alloy, can be expected for many years to come. The present would, therefore, seem to be the most desirable period for this government to engage in the preliminary negotiations necessary for the establishment of a common unit of value among all commercial nations. But the committee can make no recommendation of any specific measure beyond the resolution herewith submitted. Their conviction is clear that international uniformity is of the utmost importance for the convenience of our external trade, and of our general intercourse with foreign nations; and that at this time, especially, it is the duty of the government to prosecute with energy its efforts to effect an agreement with the leading nations of Europe on this subject. The consent of the United States, of England and of France, would necessarily ultimate in the consent of all commercial nations.

WEIGHTS AND MEASURES.

Upon the other branch of the subjects with the examination of which this committee is charged they are prepared to report more definitively.

The whole history of our revolutionary confederation, and of the constitutional government of the United States, has been a continuous acknowledgment of the perplexities arising from the diversity of weights and measures throughout their jurisdiction, and of the great desirableness of a uniform and a decimal system. The articles of confederation embraced the following clause:

"The United States, in Congress assembled, shall have the sole and exclusive right and power of regulating the alloy and value of coin struck by their own authority, or by that of the respective States, fixing the standard of weights and measures throughout the United States."

This power was transferred to Congress by the Constitution of the United States in the following language: "Congress shall have power * * * to coin money, regulate the value thereof, and of foreign coin, and fix the standard of weights and measures."

The first President, Washington, in his message to the first Congress assembled under the Constitution, brought the subject before Congress in the following language:

"Uniformity in the currency, weights and measures of the United States is an object of great importance, and will, I am persuaded, be duly attended to."

He again called the attention of Congress to it in his message of December, 1790; and again, in his opening address at the first session of the second Congress, he said:

"A uniformity in the weights and measures of the country is among the important objects submitted to you by the Constitution, and if it can be derived from a standard at once invariable and universal, must be no less honorable to the public councils than conducive to the public convenience."

In accordance with the President's first recommendation, the House of Representatives, on the 15th January, 1790—

"*Ordered*, That it be referred to the Secretary of State to prepare and report to this house, in like manner, a proper plan or plans for establishing uniformity in the currency, weights, and measures of the United States."

On the 15th of July of that year the House of Representatives received from the Secretary of State (Mr. Jefferson) his report of the proper plan for establishing the desired uniformity, as requested by the House.

In this elaborate report the Secretary proposed "that the standard of measure be a uniform, cylindrical rod of iron of such length as, in latitude 45°, in the level of the ocean, and in a cellar, or other place, the temperature of which does not vary through the year, shall perform its vibrations in uniform and equal arcs, in one second of mean time."

Starting from this standard, he proposes two distinct plans for the consideration of the House, that they might at their will adopt the one or the other exclusively, or the one for the present, and the other for the future time when the public mind may be supposed to have become familiarized to it.

The first plan was *to define and render uniform and stable the existing system;* to make the foot to bear a definite ratio to the standard pendulum rod; to reduce the dry and liquid measures to corresponding capacities by establishing a single gallon of 270 cubic inches, and a bushel to be equal to eight (8) gallons, or 2,160 inches—that is, to one and one-fourth cubic feet; to make the ounce to be the weight of one thousandth part of a cubic foot of water; to retain the more known terms of the two kinds of weights in use, reduced to one series; and to express the quantity of pure silver in the dollar in parts of the weights so defined.

The second plan was to reduce "every branch to the same decimal ratio already established in coins, and thus bring the calculation of the principal affairs of life within the arithmetic of every man who can multiply and divide plain numbers."

Except in the length of the fundamental unit, and in the nomenclature, this second plan was essentially that of the metrical system of France. A fifth part of the standard rod which has been described was taken for the foot, and was proposed for the principal unit. Its length would be about one-quarter of an inch shorter than the foot in common use.

The foot was divided into 10 inches; the inch into 10 lines; the line into 10 points; 10 feet made a decad; 10 decads made a rood; 10 roods a furlong; and 10 furlongs a mile.

The cubic foot was to be the bushel, and was to be multiplied and divided decimally for the several units of dry and liquid measure. The weight of a cubic foot of water divided decimally furnished weights. By a very slight change the silver dollar would weigh an ounce in this new series.

These two plans were sharply opposed to each other, and it was to be expected that the desire for a decimal division, and symmetry of system, on the one hand, and the reluctance to make a violent change on the other, should elicit no little discussion.

After the preparation of his report, and before its communication to the House, Mr. Jefferson received the news that propositions had been made in the national assembly of France, and in Parliament, which looked to the creation and establishment of a uniform international system of weights and measures. The movement in the former body resulted in the formation of the present metrical system of France.

This report was communicated to the Senate in December of that year, and Senators Izard, Monroe, Langdon, and Schuyler were appointed a committee to take it into consideration. That committee reported on the 1st of March, 1791, that, "as a proposition has been made to the national assembly of France for obtaining a standard of measure which shall be invariable, and communicable to all nations, and at all times; as a similar proposition has been submitted to the British Parliament in their last session; as the avowed object of these is to introduce an uniformity in the measures and weights of the commercial nations; as a coincidence of regulation by the government of the United States on so interesting a subject would be desirable, your committee are of opinion that it would not be eligible, at present, to introduce any alteration in the measures and weights which are now used in the United States." This report was adopted.

The second Congress, which met for its first session at Philadelphia, in October, 1791, had the subject again urged upon its attention by the third appeal of

President Washington in his opening address. A week later the Senate appointed a committee, consisting of very nearly the same members as the committee of the preceding Congress, to take into consideration the subject of weights and measures, and report thereon.

The committee reported on the 4th of April, 1792, recommending the adoption of the second plan proposed by Mr. Jefferson, which was an entirely deci mal and symmetrical system.

The consideration of the report was deferred until the next session of Congress, and finally referred to a special committee, and their report was not finally disposed of.

During the second session of the third Congress the President received from the French envoy a communication describing the newly adopted metric system of France, together with copies of the provisional meter and kilogram. This communication was sent to Congress on the 8th of January, 1795.

During the first session of the fourth Congress this communication of the French envoy and the report of Mr. Jefferson were referred to a committee of the House, which reported on the 12th of April, 1796. The committee assumed that all measures of surface, capacity, and weight should be regulated by measures of length; that the standard units of length and weight should not differ in a sensible degree from the present foot and pound, and that the standards should be referable to some uniform principle in nature if it can be made to appear that reference may be had to such a measure with sufficient certainty of uniformity in the result of different experiments, and without much time, trouble, and expense in making them.

They propose, therefore, that experiments be undertaken for determining the length of the proposed pendulum rod, and that from this should be derived the standard foot and standard pound. While they suggest four modes for dividing the weights, and indicate their decided preference for the decimal divisions, they avoid the vexed question of the mode of division of the foot, and also the kindred one of the contents of the bushel and gallon.

A bill to provide for the experiments passed the House, but on the third reading in the Senate was postponed to the next session, and so lost.

During the next twenty years three or four committees were appointed to examine the subject and report, but no action resulted therefrom. It was not until after the close of the war of 1812 that serious consideration of it was again resumed.

The difficulties of the questions remained. It was still uncertain whether the metrical system would eventually succeed even in France. It does not appear, therefore, that the adoption of that system was urged as a settlement. We could not be expected to give up our old measures without a resulting improvement that should be permanent. On the other hand, to divide our units decimally would destroy uniformity with England, with the unpleasant prospect before us of a second change if another decimal system should become elsewhere universal.

In his annual message, sent to the fourteenth Congress at its second session, (December 3, 1816,) President Madison urges the subject upon their attention in the following language:

"Congress will call to mind that no adequate provision has yet been made for the uniformity of weights and measures, also contemplated by the Constitution. The great utility of a standard fixed in its nature, and founded on the easy rule of decimal proportions, is sufficiently obvious. It led the government at an early stage to preparatory steps for introducing it; and a completion of the work will be a just title to the public gratitude."

As a consequence of this decided expression of the President, the Senate, just before the close of the session, (March 3, 1817,) referred it to the Secretary of State to prepare and report to them "A statement relative to the regulations and standards for weights and measures in the several States, and relative to proceedings in foreign countries for establishing uniformity in weights and measures, together with such propositions relative thereto as may be proper to be adopted in the United States." Similar action more than two years later was taken by the House of Representatives.

Without waiting for the reply of the Secretary of State, a committee of the House, on the 25th of January, 1819, presented a report on the subject. After speaking of the difficulty of introducing the new system, the committee recommends, in effect, the first plan proposed by Mr. Jefferson; also, that standards conformed to those in most common use among us should be accurately made and carefully preserved at the seat of government, and that correct models should be placed in the different districts of the country. Resolutions providing for the establishment of a commission to execute this plan, and defining the duties of such commission, were reported to the House by the committee.

The report of Hon. John Quincy Adams, Secretary of State, to whom the subject had been referred in March, 1817, was transmitted to the Senate on the 22d February, 1821. The extent of the ground covered by this learned report, and its elaborate character, satisfactorily explain why four years had been employed in its preparation. It considers, successively, the origin of measures and weights in the earlier necessities of savage life, their modifications with human progress, and, by positive law, the Hebrew, the Greek and the Roman meterology; he then adds:

"Among the nations of modern Europe there are two who, by their genius, their learning, their industry, and their ardent and successful cultivation of the arts and sciences, are scarcely less distinguished than the Hebrews, from whom they have received most of their religious, or the Greeks, from whom they have received many of their civil and political institutions. From these two nations the inhabitants of these United States are chiefly descended, and from one of them we have all our existing weights and measures. Both of them, for a series of ages, have been engaged in the pursuit of a uniform system of weights and measures. To this the wishes of their philanthropists, the hopes of their patriots, the researches of their philosophers, and the energy of their legislators have been aiming with efforts so stupendous, and with perseverance so untiring, that to any person who shall examine them it may well be a subject of astonishment to find that they are both yet entangled in the pursuit at this hour, and that it may well be doubted whether all their latest and greatest exertions have not hitherto tended to increase diversity, instead of producing uniformity."

This leads to an elaborate historical description of the English and French systems of weights and measures, together with a brief summary of the earlier discussions of the subject in this country.

The importance of uniformity between the United States and England is recognized and urged.

On the other hand, Mr. Adams was no less strongly impressed with the immense advantage of the metrical system of France. It was at that day still a question whether it would establish itself, exclusively, even in its native land, yet the hopes which it excited led the Secretary to say, with the admiration of a poet and the fervor of a prophet: "This system approaches to the ideal perfection of uniformity applied to weights and measures, and whether destined to succeed or doomed to fail, will shed unfading glory upon the age in which it was conceived, and upon the nation by which its execution was attempted, and has been in part achieved. In the progress of its establishment there, it has been often brought in conflict with the laws of physical and moral nature, with the impenetrability of matter, and with the habits, passions, prejudices and necessities of man. It has undergone various important modifications. It must undoubtedly still submit to others before it can look for universal adoption. But if man upon earth be an improvable being, if that universal peace which was the object of a Savior's mission, which is the desire of the philosopher, the longing of the philanthropist, the trembling hope of the Christian, is a blessing to which the futurity of mortal man has a claim of more than mortal promise; if the spirit of evil is, before the final consummation of things, to be cast down from his dominion over men and bound in the chains of a thousand years, the foretaste here of man's eternal felicity, then this system of instruments, to accomplish all the changes of social and friendly commerce, will furnish the links of sympathy between the inhabitants of the most distant regions; the meter will surround the globe in use as well as in multiplied extension, and one language of weights and measures will be spoken from the equator to the poles."

After an analysis and contract of the respective advantages and disadvantages of the English and French weights and measures, so far as the advantages or disadvantages could be derived from theory, and the very imperfect experience of. the French up to that time, Mr. Adams adds:

"These views are presented as leading to the conclusion that, as final and universal uniformity of weights and measures is the common desideratum of all civilized nations, as France has formed, and has for her own use established, a system adapted by the highest efforts of human science, ingenuity and skill, to the common purposes of all; as this system is yet new, imperfect, susceptible of great improvements, and struggling for existence even in the country which gave it birth, as its universal establishment would be a universal blessing, and as, if ever effected, it can only be by consent, and not by force, in which the energies of opinion must precede those of legislation, it would be worthy of the dignity of the Congress of the United States to consult the opinions of all the civilized nations with whom they have a friendly intercourse, to ascertain, with the utmost attainable accuracy, the existing state of their respective weights and measures, to take up and pursue with steady, persevering, but always temperate and discreet exertions, the idea conceived, and thus far executed, by France, and to co-operate with her to the final and universal establishment of her system." * * *

"In contemplating so great but so beneficial a change as the ultimate object of the proposal now submitted to the consideration of Congress, it is supposed to be most congenial to the end to attempt no present change whatever in our existing weights and measures, to let the standards remain precisely as they are, and to confine the proceedings of Congress at this time to authorizing the Executive to open these communications with the European nations where we have accredited ministers and agents, and to such declaratory enactments and regulations as may secure a more perfect uniformity in the weights and measures now in use throughout the Union."

After giving statements of the laws in force in the several States, the report concludes by submitting to Congress a plan consisting of two parts, the principles of which were:

1. To fix the standard with the partial uniformity of which it is susceptible, for the present excluding all innovation.

2. To consult with foreign nations for the future and ultimate establishment of universal and permanent uniformity.

"All trifling and partial attempts of change in our existing system, it is hoped, will be steadily discountenanced and rejected by Congress, not only as unworthy of the high and solemn importance of the subject, but as impracticable to the purpose of uniformity, and as inevitably tending to the reverse, to increased diversity, to inextricable confusion."

Congress has heretofore authorized the construction of standards of the common measures of length, weight, and capacity, and their distribution to the several States, as well as to the custom-houses and certain departments of the public service, but has hitherto failed to take a decisive step in advance.

In the mean time, the separate action of foreign governments, as will hereaft e be shown, has produced the results which the Secretary sought by his proposition for concurrent action. The desire he expressed for the concurrence of the British government especially is now realized in the initiatory steps taken by Parliament in the authorized adoption of the metric system.

For this and other prior reasons the second part of Mr. Adams's plan has not been effectively prosecuted. Its objects have not, however, been forgotten, and have occupied, during the last ten years more especially, the serious attention of the people and the government. Resolutions of State legislatures, petitions from scientific and other organized societies, recommendations from executive officers, and direct action of Congress—these all indicate a dissatisfaction with the present defective system of our weights and measures and an earnest desire for a decimal system common to all nations.

In his annual report of December 9, 1847, the Secretary of the Treasury (Hon. R. J. Walker) commends the subject to the attention of Congress. He says:

"Coins, as well as weights and measures, for the benefit of all nations, ought to be uniform throughout the world; and if our decimal system of coinage should be more simple and perfect than that of any other nation, it ought to be, and ultimately will be, adopted, and lead as far as practicable to the introduction of the decimal system of weights and measures, or at least its simplification, so that ultimately the coin and the weights and measures may be simple and uniform throughout the world."

A few months later the superintendent of weights and measures, Professor Bache, in his report, urges attention to the subject. (Ex. Doc. 84, 30th Cong., 1st sess., July 30, 1848:)

"No one who has discussed the subject of weights and measures in our country has considered the present arrangement as an enduring one. It has grown up with the growth of European society, and is deficient in simplicity and in system. The labor which is expended in mastering the complex denominations of weights and measures is labor lost. Every purpose for which weights and measures are employed can be answered by a simple and connected arrangement. * * * * * * *

"In our own country the present arrangement of weights and measures has been considered by the great men who have written upon it as temporary."

After speaking of the two parts of Mr. Adams's plan, and asking whether the time has now come for urging the measures involved in the second part, he says:

"The present time seems especially to invite an effort of this kind. In England the subject of weights and measures is under consideration by a commission, and on the continent the new relations of States hitherto separated appear to be favorable to this object. Such changes could readily be effected by suitable means in one generation by introducing the new measures through the elementary schools.

* * * * * * * * *

"I am of opinion that the present weights and measures, whether declared to be provisional or not, will prove to be really so in the progress of our Union, and that arrangements more worthy to be called a system will one day prevail."

In a subsequent report he says:

"In Holland the new weights and measures were introduced through the schools. The children of the country becoming familiar with them in the primary schools, seeing the actual material standards of length, capacity, and weight at frequent and stated times in early youth, and retaining that familiarity as they passed into the higher schools, would be readily prepared for their universal use when reaching mature life. But the old material standards must disappear, and not, as in our coinage, be tolerated by usage alongside of the lawful standards, destroying what Mr. Adams has so well called the uniformity of fact.

* * * * * * * * *

"Coming into the charge of an unfinished work, I conformed, as far as I could, to the plans already in part executed by my predecessor, Mr. Hassler, as I could co-operate heartily in the endeavor to produce that uniformity of fact which was the basis of the system. I have not failed, from time to time, to press forward the second part of this established system, namely, the endeavor at universal uniformity.

* * * * * * * * *

"The first part of Mr. Adams's plan has (as far as legal standards are concerned, and in a great degree) been accomplished; but the second part, that which recommends the consultation with foreign nations for the ultimate establishment of universal uniformity remains yet to be acted on."

* * * * * * * *

"By reference to the interesting account of the metrical system in the letter of Mr. Silbermann, it will be seen that it has extended widely beyond the boundaries of France, and has been adopted by law in Spain, Belgium, Greece, Holland, Lombardy, Poland, and Switzerland, in Europe; and Chili, Colombia, and Mexico, on this continent."

* * * * * * * *

"Has not the time arrived, in the general progress of commercial and international intercourse, and the rapid advance of our own country in science, wealth, and power, when her voice should be heard in an important matter like this? Should not Congress make the proposition to all nations to meet by their representatives, and consult for the purpose of establishing permanent and universal uniformity of weights and measures? Such action could not fail to meet with a response due to the greatness of the subject, and, if the great object be attained, to lead to results productive of vast and lasting benefit to the human race."

The legislature of New Hampshire, by joint resolution, approved June 28, 1859, requested their senators and representatives to urge upon Congress the adoption of a decimal system. The legislature of Maine, March 20, 1860, by joint resolution, expressed in still more decided language their desire for a uniform international decimal system of weights and measures and coins.

The legislature of Connecticut by resolution (1861) seconded this action of Maine. In June, 1864, they further recommended to the proper school officers to provide for teaching the metrical system in all schools of the State.

The Secretary of the Treasury, (now Chief Justice of the United States,) in his annual report, December 9, 1861, again brought the subject to the attention of Congress.

"The Secretary desires to avail himself of this opportunity to invite the attention of Congress to the importance of a uniform system, and a uniform nomenclature of weights, measures, and coins, to the commerce of the world, in which the United States already so largely shares. The wisest of our statesmen have regarded the attainment of this end so desirable in itself as by no means impossible. The combination of the decimal system with appropriate denominations in a scheme of weights, measures, and coins for the international uses of commerce, leaving, if needs be, the separate systems of nations untouched, is certainly not beyond the reach of the daring genius and patient endeavor which gave the steam-engine and the telegraph to the service of mankind. The Secretary respectfully suggests the expediency of a small appropriation to be used in promoting interchange of opinions between intelligent persons of our own and foreign countries on this subject."

In May, 1863, an international postal congress was held at Paris, at the suggestion of the government of the United States, in which nearly all the European and some of the American governments were represented. Among the resolutions adopted at that congress were the following:

"SEC. 7. The rates upon international correspondence shall be established according to the same scale of weight in all countries.

"SEC. 8. The metrical decimal system, being that which best satisfies the demands of the postal service, shall be adopted for international postal relations, to the exclusion of every other system.

"SEC. 9. The single rate upon international letters shall be applied to each standard weight of fifteen grams, or fractional part of it."

At that congress, representing nations having many different systems of weights and measures, the expression in favor of the metric system was unanimous.

In the autumn of 1863 an international statistical congress was held at Berlin, and at the instance of the Prussian government the Secretary of State appointed a commissioner to represent the United States therein. His report of the transactions of the congress was transmitted to the Senate on the 18th of June, 1864. All, or nearly all, of the nations of Europe were represented. The subject of a uniform international system of coinage, weights, and measures was presented upon the report of a numerous committee, which had been appointed at the previous meeting held at London three years before. This committee contained representatives of fourteen different nations. Its report was transmitted to Congress and published. It contains detailed information as to the weights, measures, and coins of the European and a portion of the American nations. The statistical congress, after discussion, resolved that the adoption of the same measures in international commerce is of the highest importance, and that the metric system is the most convenient of all that can be recommended for international measures.

In the same year, (1863,) by request of the Secretary of the Treasury, the National Academy of Sciences appointed a committee to consider and report upon the subjects embraced within the jurisdiction of this committee. After patient investigation and deliberate discussion that committee made the following report, which was adopted by the Academy with almost entire unanimity:

"*Report of the Committee on Weights, Measures, and Coinage, to the National Academy of Sciences, January,* 1866.

"The committee are in favor of adopting, ultimately, a decimal system; and in their opinion, the metrical system of weights and measures, though not without defects, is, all things considered, the best in use. The committee therefore suggest that the academy recommend to Congress to authorize and encourage by law the introduction and use of the metrical system of weights and measures; and that with a view to familiarize the people with the system, the academy recommend that provision be made by law for the immediate manufacture and distribution to the custom-houses and States of metrical standards of weights and measures; to introduce the system into the post offices by making a single letter weigh fifteen grammes instead of fourteen and seventeen-hundredths or half an ounce; and to cause the new cent and two-cent pieces to be so coined that they shall weigh, respectively, five and ten grammes, and that their diameters shall be made to bear a determinate and simple ratio to the metrical unit of length."

In concluding this review of the agitation of and action upon these questions in the United States, it only remains to add that the House of Representatives of the 38th Congress, at its first session, established, by an amendment of its rules, a standing committee to take jurisdiction of this great reform. As efforts to carry that reform into effect had hitherto been spasmodic rather than consecutive, it was thought proper thereafter to crystallize them through the action of a permanent committee, before whom they should perpetually reappear until this conceded great desideratum should become an accomplished fact.

But while the United States was the first to move in the direction of a decimal system resting upon a natural and universally attainable standard, the effect of the delay of this government, with a view to harmony in action with England, has been to render it possible that the United States will be among the last in the column of nations to take this great step in civilization.

Our predecessors of the era of Mr. Adams found the interests of this country much more dependent upon England than they are at this day. England herself was less subject at that time to the effect of foreign influence than at present. The failure of these two governments to unite upon a system resting upon a standard of their own, at a time when France stood alone for the metric system, has been fatal to the adoption of the arbitrary system of those countries by other nations. Convinced of its imperfections, no effort was made to introduce it into other countries, and any modification of it with a view to its improvement would only have created an additional system to those already in use in the world, without having in any of its features a superiority over the metric system. In the mean time the simple order, beauty, and convenience of the metric system has so commended it to universal acceptance that it has already been adopted exclusively or permissively by nearly all the nations of christendom.

In France, Spain, Belgium, and Portugal, it has been established to the exclusion of other weights and measures. In Holland other weights are allowed in compounding medicines only. Sardinia and Lombardy have long possessed the system, and it has now been extended to the whole of Italy. Greece has introduced it with some modifications. In Austria, and most of the other German States, the half kilogram has been for some time a common unit of weight in the custom-houses, and on railways. During the past year your committee are informed that delegates of all the German States, at a meeting at Frankfort-on-the-Main, signed a convention agreeing to introduce into the several States systems of which the meter should be the basis. Prussia, which had previously withheld assent, thus appears to join in the common movement. Switzerland will necessarily follow Germany, and already has units that are aliquot parts of the meter and the kilogram. The King of Sweden and Norway has appointed a commission to consider and report on the best mode of introducing the metric system among his subjects. Denmark may be expected to follow the recommendation of the Scandinavian convention that advised it. We have the assurance of M. Kupfer, the distinguished superintendent of weights and measures of the Russian empire, that if England should adopt the metric system, Russia will also adopt it.

The system has also made great progress among the States upon this continent.

Six years since it was adopted by the Mexican republic, and its use decreed at once in the public offices, and after a certain period in private contracts. This period expired about the time of the imperial invasion under which that republic is now suffering. It was introduced into Chili in 1848, and is compulsory from the 1st of June, 1865. In the United States of Colombia, and in Venezuela, it has been in use along with other weights and measures since 1853. In Brazil the meter is used for cloth measure, and the liter for wine measure. In Equador the system was decreed to come into full operation on the 15th of October next. In Guatemala, San Salvador, and the Argentine Republic, it is in partial use among the people.

The action of England is, however, of greater importance to us, owing to our close relations with her, and with her colonies, by a common language, by our large commerce, and what is, perhaps, more pertinent to this question, by common weights and common measures.

On the 8th of April, 1862, the House of Commons appointed a select committee of fifteen members to "consider the practicability of adopting a simple and uniform system of weights and measures, with a view not only to the benefit of internal trade, but to facilitate trade and intercourse with foreign countries." The committee examined thirty-nine witnesses, among whom were nine from foreign countries in which the metric system was in force. They were generally men of distinguished intelligence, who were attending the Industrial Exhibition as commissioners from their respective countries. The list of witnesses included seven merchants, six civil engineers and architects, ten professors and teachers, two manufacturers, four actuaries and accountants, the astronomer royal, the master of the mint, and the secretary of the post office. That committee appear to have been unanimous in recommending the introduction of the metrical system into Great Britain.

On the 13th of May, 1863, a bill was prepared and brought in by members of that committee, by the terms of which the metric system of weights and measures was introduced into Great Britain, and its use by the people made compulsory after three years. This bill was passed by the House of Commons by a large majority, but does not appear to have been acted on by the House of Lords. At the next session (February, 1864) a bill was introduced by the same gentlemen which changed its purport from a compulsory to a permissive measure.

This bill passed the House of Commons on the 29th of June, the House of Lords on the 21st of July, and became a law. The vote of the House of Commons approving a compulsory measure, and the subsequent enactment of a permissive law, must be regarded as evincing a deliberate intention to introduce the metric system into England, and as giving up any purpose of creating a separate system founded upon the yard, the foot, or the inch; and as paving the way for the ultimate exclusive adoption of the metric scheme.

The general consent of so many nations, highly enlightened, and deeply interested in the promotion of trade, and in popular progress, affords in itself an argument almost conclusive for, first, uniformity; second, decimalization; and, third, the metric plan.

But habit makes us so submissive even to constant inconvenience that your committee submit herewith three tables showing the perplexities and embarrassments involved in our customary weights and measures, in every effort for their mutual conversion, and even in all efforts of the memory to retain the relations of their several parts. They multiply most seriously the arithmetical rules required, embarrass mathematical calculations, clog the accounts of trade, increase the labors of teachers and scholars alike in our schools, absorb in their acquisition a great portion of the time which would be more usefully applied to other studies, and necessarily appreciate the cost of a common business education. With a decimal system all the ordinary transactions of popular trade could be computed by any person familiar with the simplest relations of numbers, and without pencil or paper to aid the mind, now embarrassed by their complexity. But with the actual system in use, in the table of lengths we ascend by the factors 12, 3, $5\frac{1}{2}$, 40, 8, and 3; or else by $7\frac{23}{25}$, 25, 4, and 80.

In weights we have three series, nearly distinct—Avoirdupois, Troy, and Apothecary's. The only common unit is the grain. In the first, we ascend from grain by the factors $27\frac{11}{32}$, 16, 16, 25 or 28, 4, and 20; in the second the factors are 24, 20, and 12; in the third, 20, 3, 8, and 12.

In measures of capacity there are simple relations between the several liquid measures, as well as between the dry measures, and also the cubic measures; yet, in comparing the measures of the three different series, there are no useful relations whatever.

The accompanying tables exhibit to the eye this want of system. They give the number by which it is necessary to multiply, or divide, in order to reduce one denomination to another. These factors, when fractional, are reduced to their lowest terms.

I. *Table of usual lengths, exhibiting the number of units of each denomination contained in each larger denomination.*

	Lines.	Inches.	Links.	Feet.	Yards.	Fathoms.	Rods.	Chains.	Furlongs.	Miles.	Leagues.
The inch contains	12	1									
The link contains	$\frac{2376}{25}$	$\frac{198}{25}$	1								
The foot contains	144	12	$\frac{50}{33}$	1							
The yard contains	432	36	$\frac{50}{11}$	3	1						
The fathom contains	864	72	$\frac{100}{11}$	6	2	1					
The rod contains	2376	198	25	$\frac{33}{2}$	$\frac{11}{2}$	$\frac{11}{4}$	1				
The chain contains	9504	792	100	66	22	11	4	1			
The furlong contains	95040	7920	1000	660	220	110	40	10	1		
The mile contains	760320	63360	8000	5280	1760	880	320	80	8	1	
The league contains	2280960	190080	24000	15840	5280	2640	960	240	24	3	1

II. *Table of weights in use, exhibiting the number of units of each denomination contained in each larger denomination.*

	Grains.	Scruples.	Penny-weights.	Avoir. drachms.	Apoth. drachms.	Avoir. ounces.	Troy ounces.	Troy pounds.	Avoir. pounds.	Short ton.	Long ton.
The scruple contains	20	1									
The pennyweight contains	24	$\frac{6}{5}$	1								
The avoir. drachm contains	$\frac{875}{32}$	$\frac{175}{128}$	$\frac{875}{768}$	1							
The apoth. drachm contains	60	3	$\frac{5}{2}$	$\frac{384}{175}$	1						
The avoir. ounce contains	$\frac{875}{2}$	$\frac{175}{8}$	$\frac{875}{48}$	16	$\frac{175}{24}$	1					
The Troy ounce contains	480	24	20	$\frac{3072}{175}$	8	$\frac{192}{175}$	1				
The Troy pound contains	5760	288	240	$\frac{36864}{175}$	96	$\frac{2304}{175}$	12	1			
The avoir. pound contains	7000	350	$\frac{875}{3}$	256	$\frac{350}{3}$	16	$\frac{175}{12}$	$\frac{175}{144}$	1		
The short ton contains	14000000	700000	$\frac{1750000}{3}$	512000	$\frac{700000}{3}$	32000	$\frac{87500}{3}$	$\frac{21875}{9}$	2000	1	
The long ton contains	15680000	784000	$\frac{1960000}{3}$	573440	$\frac{784000}{3}$	35840	32000	$\frac{8000}{3}$	2240	$\frac{28}{25}$	1

III. *Table of usual measures of capacity, exhibiting the number of units of each denomination contained in each larger denomination.*

	Fluid drachms.	Cubic inches.	Fluid ounces.	Gills.	Pints.	Fluid quarts.	Dry quarts.	Gallons.	Pecks.	Cubic feet.	Bushels.	Cubic yards.
The cubic inch contains ...	$\frac{1024}{231}$	1										
The fluid ounce contains ...	8	$\frac{231}{128}$	1									
The gill contains	32	$\frac{231}{32}$	4	1								
The pint contains	128	$\frac{231}{8}$	16	4	1							
The quart contains	256	$\frac{231}{4}$	32	8	2	1						
The dry quart contains	$\frac{1720336}{5775}$	$\frac{107521}{1600}$	$\frac{215042}{5775}$	$\frac{107521}{11550}$	$\frac{107521}{46200}$	$\frac{107521}{92400}$	1					
The gallon contains	1024	231	128	32	8	4	$\frac{369600}{107521}$	1				
The peck contains	$\frac{13762688}{5775}$	$\frac{107521}{200}$	$\frac{1720336}{5775}$	$\frac{430084}{5775}$	$\frac{107521}{5775}$	$\frac{107521}{11550}$	8	$\frac{107521}{46200}$	1			
The cubic foot contains	$\frac{590224}{77}$	1728	$\frac{73728}{77}$	$\frac{18432}{77}$	$\frac{4608}{77}$	$\frac{2304}{77}$	$\frac{691200}{107521}$	$\frac{576}{77}$	$\frac{345600}{107521}$	1		
The bushel contains	$\frac{65050752}{5775}$	$\frac{107521}{50}$	$\frac{6881344}{5775}$	$\frac{1720336}{5775}$	$\frac{430084}{5775}$	$\frac{215042}{5775}$	32	$\frac{107521}{11550}$	4	$\frac{107521}{86400}$	1	
The cubic yard contains ...	$\frac{15925248}{77}$	46656	$\frac{1990656}{77}$	$\frac{497664}{77}$	$\frac{124416}{77}$	$\frac{62208}{77}$	$\frac{18662400}{107521}$	$\frac{15552}{77}$	$\frac{9331200}{107521}$	27	$\frac{2332800}{107521}$	1

These tables indicate that none but professional persons can be expected to master and retain their knowledge of the arithmetical intricacies of our present scheme, as taught in schools and used in practical life. If, to those denominations there mentioned, we should add nails, ells, barleycorns, the two quarters, the two cwts., the ale measures, and the various barrels, pipes, and hogsheads, the lists of difficulties would be formidably increased.

In marked contrast with this is

THE METRIC SYSTEM.

It is orderly, simple, and perfectly harmonious, having useful relations between all its parts. It is based on the METER, which is its principal and only arbitrary unit. The meter is a measure of length, and was intended to be, and is very nearly, one ten-millionth of the distance on the earth's surface from the equator to the pole. It is 39.37 inches, very nearly.

The *are* is a surface equal to a square whose side is 10 meters. It is nearly four square rods.

The *liter* is the unit for measuring capacity, and is equal to the contents of a cube whose edge is a tenth part of the meter. It is a little more than a wine quart.

The *gram* is the unit of weight, and is the weight of a cube of water, each edge of the cube being one one-hundredth of the meter. It is equal to 15,432 grains.

The *stere* is a cubic meter.

Each of these units is divided decimally, and larger units are formed by multiples of 10, 100, &c. The successive multiples are designated by the prefixes deka, hecto, kilo, and myria; the subordinate parts by deci, centi, and milli; each having its own numerical significance.

Scheme of the weights and measures of the metric system.

Ratios.	Lengths.	Surfaces.	Volumes.	Weights.
1000000				Millier, or tonneau.
100000				Quintal.
10000	Myriameter.			Myriagram.
1000	Kilometer.		Kiloliter, or stere.	Kilogram, or kilo.
100	Hectometer.	Hectare	Hectoliter.	Hectogram.
10	Dekameter.		Dekaliter.	Dekagram.
1	METER.	ARE.	LITER.	GRAM.
$\frac{1}{10}$	Decimeter.		Deciliter.	Decigram.
$\frac{1}{100}$	Centimeter.	Centare.	Centiliter.	Centigram.
$\frac{1}{1000}$	Millimeter.		Milliliter.	Milligram.

Scale representing a portion of the meter divided into centimeters and millimeters, together with a six-inch scale divided into eighths of an inch; one inch contains 25.4 millimeters.

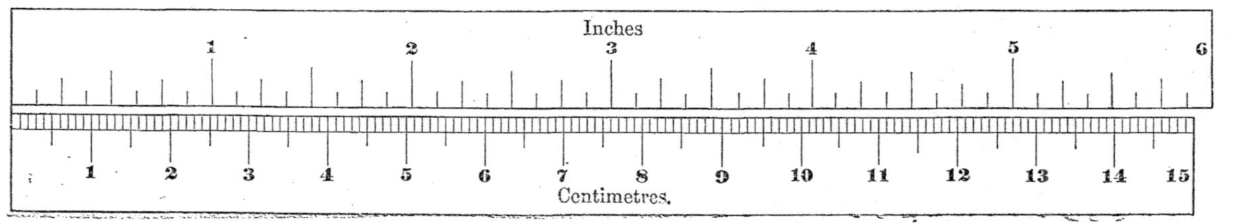

The nomenclature, simple as it is in theory, and designed from its origin to be universal, can only become familiar by use. Like all strange words, these will become familiar by custom, and obtain popular abbreviations. A system which has incorporated with itself so many different series of weights, and such a nomenclature as "scruples," "pennyweights," "avoirdupois," and with no invariable component word, can hardly protest against a nomenclature whose leading characteristic is a short component word, with a prefix signifying number. We are already familiar with *thermometer*, *barometer*, *diameter*, *gasometer*, &c., with *telegram*, *monogram*, &c.—words formed in the same manner.

After considering every argument for a change of nomenclature, your committee have come to the conclusion that any attempt to conform it to that in present use would lead to confusion of weights and measures; would violate the easily-learned order and simplicity of metric denomination, and would seriously interfere with that universality of system so essential to international and commercial convenience.

When it is remembered that of the value of our exports and imports in the year ending June 30, 1860, in all $762,000,000, the amount of near $700,000,000 was with nations and their dependencies that have now authorized, or taken the preliminary steps to authorize, the metric system, even denominational uniformity for the use of accountants in such vast transactions assumes an important significance. In words of such universal employment each word should represent the identical thing intended, and no other, and the law of association familiarizes it.

Table of the commerce of the United States for the year ending June 30, 1860, *exhibiting the value of the exports to and imports from each foreign country (including its colonies) in which the metric system is entirely or partially adopted or is in process of adoption; and also the exports to and imports from all other nations.*

METRIC NATIONS.

Countries.	Exports.	Imports.
Sweden, Norway, and Swedish West Indies	$1,516,345	$532,984
Hamburg, Bremen, and other German ports	18,427,958	18,498,607
Holland and Dutch colonies	4,867,738	4,501,306
Belgium	4,559,748	2,558,873
Great Britain and colonies	238,887,117	177,913,585
France and colonies	63,050,187	43,409,627
Spain and colonies	21,165,794	44,492,314
Portugal and colonies	402,303	266,440
Italy	5,073,375	4,734,518
Austria	1,038,904	732,645
Greece	....	71,754
Mexico	5,354,073	6,935,872
Central America	149,698	331,258
New Granada	1,795,499	3,843,568
Venezuela	1,147,900	2,883,464
Brazil	6,280,255	21,214,803
Argentine republic	999,708	4,020,848
Chili	3,268,673	2,072,912
Equador	19,545	....
Total	378,004,820	339,015,378

NON-METRIC NATIONS.

Countries.	Exports.	Imports.
Russia and possessions	$2,833,325	$1,557,868
Prussia	....	36,464
Ionian republic	....	62,897
Denmark and Danish West Indies	1,328,548	216,925
Turkey	849,768	970,250
Egypt	36,420	71,709
African ports	2,370,543	1,755,916
Hayti	2,673,682	2,062,723
San Domingo	169,300	283,098
Uruguay	789,358	908,750
Peru	987,672	308,452
Sandwich Islands	747,462	367,859
Other Pacific islands	65,274	112,401
Japan	138,774	55,091
China	8,906,118	13,566,587
Other ports in Asia	108,969	49,634
Total	22,005,213	22,386,624
Whale fisheries and unknown	112,263	764,252

Your committee unanimously recommend the passage of the bills and joint resolutions appended to this report. They were not prepared to go, at this time, beyond this stage of progress in the proposed reform. The metric system is already used in some arts and trades in this country, and is especially adapted to the wants of others. Some of its measures are already manufactured at Bangor, in Maine, to meet an existing demand at home and abroad. The manufacturers of the well-known Fairbanks scales state: "For many years we have had a large export demand for our scales with French weights, and the demand and sale is constantly increasing." Its minute and exact divisions specially adapt it to the use of chemists, apothecaries, the finer operations of the artisan, and to all scientific objects. It has always been and is now used in the United States coast survey. Yet in some of the States, owing to the phraseology of their laws, it would be a direct violation of them to use it in the business transactions of the community. It is therefore very important to legalize its use, and give to the people, or that portion of them desiring it, the opportunity for its legal employment, while the knowledge of its characteristics will be thus diffused among men. Chambers of commerce, boards of trade, manufacturing associations, and other voluntary societies, and individuals, will be induced to consider and in their discretion to adopt its use. The interests of trade among a people so quick as ours to receive and adopt a useful novelty, will soon acquaint practical men with its convenience. When this is attained—a period, it is hoped, not distant—a further act of Congress can fix the date for its exclusive adoption as a legal system. At an earlier period it may be safely introduced into all public offices, and for government service.

In the schedule of equivalents provided in the bill, extreme scientific accuracy is not expressed. The reasons follow. The exact length of the meter in inches and the weight of the kilogram in grains can of necessity be determined only approximately. The most careful determinations of these quantities now possible are liable to minute corrections hereafter, as more numerous observations are made and better instruments are used. Instead, therefore, of aiming at an accuracy greater, perhaps, than is attainable, it is more expedient to consult the convenience of the people by using the simplest numbers possible in the schedule, and yet such as shall be in fact more nearly exact than can ever be demanded in the ordinary business of life. These numbers are to be used in schools, and in practical life millions of times as multipliers and divisors, and every unnecessary additional figure is justly objectionable.

In a popular sense of the word, however, the numbers in the schedule may be said to be exact. The length of the meter, for example, is given as 39.37 inches. The mean of the best English and the best American determinations differs from this only by about the amount by which the standard bar changes its length by a change of one degree of temperature. Such accuracy is certainly sufficient for legal purposes and for popular use.

The second measure recommended is a joint resolution, necessarily following the adoption of the leading bill, and provides for furnishing the standards which will thereby be required, to the several States.

The third proposition is a bill to authorize and provide for the use of the weight of 15 grams in the post office, in conformity with the system adopted by that department for foreign correspondence.

The fourth is a resolution looking to effective negotiation for a uniform coinage among nations.

Respectfully submitted,

JOHN A. KASSON, *Chairman.*
CHARLES H. WINFIELD.
THOMAS WILLIAMS.
HEZEKIAH S. BUNDY.
HENRY L. DAWES.

BILLS AND RESOLUTIONS ACCOMPANYING THE REPORT.

A BILL to authorize the use of the metric system of weights and measures.

Be it enacted by the Senate and House of Representatives of the United States of America in Congress assembled, That from and after the passage of this act, it shall be lawful throughout the United States of America to employ the weights and measures of the metric system; and no contract, or dealing, or pleading in any court, shall be deemed invalid, or liable to objection, because the weights or measures expressed or referred to therein are weights or measures of the metric system.

SEC. 2. *And be it further enacted,* That the tables in the schedule hereto annexed shall be recognized, in the construction of contracts, and in all legal proceedings, as establishing, in terms of the weights and measures now in use in the United States, the equivalents of the weights and measures expressed therein in terms of the metric system; and said tables may be lawfully used for computing, determining, and expressing in customary weights and measures the weights and measures of the metric system.

Measures of length.

METRIC DENOMINATIONS AND VALUES.		EQUIVALENTS IN DENOMINATIONS IN USE.
Myriameter	10,000 meters	6.2137 miles.
Kilometer	1,000 meters	0.62137 mile, or 3,280 feet and 10 inches.
Hectometer	100 meters	328 feet and one inch.
Dekameter	10 meters	393.7 inches.
Meter	1 meter	39.37 inches.
Decimeter	$\frac{1}{10}$th of a meter	3.937 inches.
Centimeter	$\frac{1}{100}$th of a meter	0.3937 inch.
Millimeter	$\frac{1}{1000}$th of a meter	0.0394 inch.

Measures of surface.

METRIC DENOMINATIONS AND VALUES.		EQUIVALENTS IN DENOMINATIONS IN USE.
Hectare	10,000 square meters	2.471 acres.
Are	100 square meters	119.6 square yards.
Centare	1 square meter	1,550 square inches.

Measures of capacity.

METRIC DENOMINATIONS AND VALUES.			EQUIVALENTS IN DENOMINATIONS IN USE.	
Names.	No. of liters.	Cubic measure.	Dry measure.	Liquid or wine measure.
Kiloliter or stere	1000	1 cubic meter	1.308 cubic yards	264.17 gallons.
Hectoliter	100	$\frac{1}{10}$ of a cubic meter	2 bus. and 3.35 pecks	26.417 gallons.
Dekaliter	10	10 cubic decimeters	9.08 quart	2.6417 gallons.
Liter	1	1 cubic decimeter	0.908 quarts	1.0567 quarts.
Deciliter	$\frac{1}{10}$	$\frac{1}{10}$ of a cubic decimeter	6.1022 cubic inches	0.845 gill.
Centiliter	$\frac{1}{100}$	10 cubic centimeters	0.6102 cubic inch	0.338 fluid ounce.
Milliliter	$\frac{1}{1000}$	1 cubic centimeters	0.061 cubic inch	0.27 fluid drachm.

Weights.

METRIC DENOMINATIONS AND VALUES.			EQUIVALENTS IN DENOMINATIONS IN USE.
Names.	Number of grams.	Weight of what quantity of water at maximum density.	Avoirdupois weight.
Millier or tonneau..	1000000	1 cubic meter	2204.6 pounds.
Quintal	100000	1 hectoliter	220.46 pounds.
Myriagram	10000	10 liters	22.046 pounds.
Kilogram or kilo...	1000	1 liter	2.2046 pounds.
Hectogram	100	1 deciliter	3.5274 ounces.
Dekagram	10	10 cubic centimeters	0.3527 ounce.
Gram	1	1 cubic centimeters	15.432 grains.
Decigram	$\frac{1}{10}$	$\frac{1}{10}$ of a cubic centimeter	1.5432 grains.
Centigram	$\frac{1}{100}$	10 cubic millimeters	0.1543 grain.
Milligram	$\frac{1}{1000}$	1 cubic millimeter	0.0154 grain.

JOINT RESOLUTION to enable the Secretary of the Treasury to furnish to each State one set of the standard weights and measures of the metric system.

Be it resolved by the Senate and House of Representatives of the United States of America in Congress assembeld, That the Secretary of the Treasury be, and he is hereby, authorized and directed to furnish to each State, to be delivered to the governor thereof, one set of the standard weights and measures of the metric system, for the use of the States respectively.

A BILL to authorize the use in post offices of weights of the denomination of grams.

Be it enacted by the Senate and House of Representatives of the United States of America in Congress assembled, That the Postmaster General be, and he is hereby, authorized and directed to furnish to the post offices exchanging mails with foreign countries, and to such other offices as he shall think expedient, postal balances denominated in grams of the metric system, and until otherwise provided by law, one-half ounce avoirdupois shall be deemed and taken for postal purposes as the equivalent of fifteen grams of the metric weights, and so adopted in progression; and the rates of postage shall be applied accordingly.

JOINT RESOLUTION to authorize the President to appoint a special commissioner to facilitate the adoption of an uniform coinage between the United States and foreign countries.

Be it resolved by the Senate and House of Representatives of the United States of America in Congress assembled, That the President be, and he is hereby, authorized to appoint a special commissioner to negotiate with foreign governments for the establishment of the common unit of money, of identical value in all commercial countries adopting the same; that all governments with which the United States hold diplomatic relations be invited to participate in the negotiations. That any plan which may be agreed upon by part of or all the representatives engaging in such negotiations be submitted to Congress for its approval before being carried into effect in the United States, and that the compensation allowed to such commissioner be the amount necessary for his actual and proper expenses incurred in the execution of his duties.

www.ingramcontent.com/pod-product-compliance
Lightning Source LLC
LaVergne TN
LVHW011141110826
845150LV00008B/2454

* 9 7 8 1 4 1 8 1 9 3 3 0 0 *